SAFE USE OF RADIOACTIVE TRACERS
IN INDUSTRIAL PROCESSES

SAFETY SERIES No. 40

SAFE USE OF RADIOACTIVE TRACERS IN INDUSTRIAL PROCESSES

SPONSORED BY THE
INTERNATIONAL ATOMIC ENERGY AGENCY
AND THE
WORLD HEALTH ORGANIZATION

INTERNATIONAL ATOMIC ENERGY AGENCY
VIENNA, 1974

THIS SAFETY SERIES WILL ALSO BE PUBLISHED IN FRENCH

SAFE USE OF RADIOACTIVE TRACERS
IN INDUSTRIAL PROCESSES
IAEA, VIENNA, 1974
STI/PUB/369

Printed by the IAEA in Austria
January 1974

FOREWORD

The techniques for using radioactive tracers in industrial processes have become widely established. Although the associated
risks to radiation workers and to the public have been small, this
use of tracers has greatly increased in recent years. At present
there are few routine applications of tracers which could lead to the
continuous introduction of radioactivity into commercial products
or directly into the environment, and this situation is likely to
continue into the foreseeable future. Although industrial applications of tracers will continue to contribute only a very small pro-
portion of the population dose relative to other activities, some guide
is deemed necessary for those who may be involved in planning and
implementing such tracer investigations as well as for those who
may have to institute regulatory functions.

In 1970 the International Atomic Energy Agency (IAEA), in
collaboration with the World Health Organization (WHO), engaged
the services of Mr. R. Cunningham (USA), Mr. B. Fries (USA) and
Mr. P. Johnson (UK) as consultants to prepare a draft manual. This
draft then formed the working paper for a panel of experts convened
by the IAEA and WHO in 1972. At this stage the International Labour
Organization (ILO), the OECD Nuclear Energy Agency (NEA) and
the Food and Agriculture Organization of the United Nations (FAO)
expressed their interest in the preparation of the final guide but
were unable to participate in the panel meeting. The revised text
that resulted from the panel's work was reviewed and put into the
present final form by Mr. Cunningham.

A number of examples of radioactive tracer applications provided by the consultants and panel members are presented in the
Annexes.

CONTENTS

1. INTRODUCTION

Radiotracers have wide application in industry and provide valuable methods for the measurement and investigation of industrial process systems. This is because the ideal tracer is one which is identical to a component of the investigated system and yet can be unambiguously measured. In addition, the radioactive tracers offer ease of measurement and high sensitivity together with absence of interference with the studied process. Well established radiotracer methods include those for flow measurement, determination of residence times, flow patterns, mixing efficiencies, ventilation efficiency studies, location of leaks, etc. Detailed descriptions of techniques and applications are to be found in the technical literature, including the published proceedings of various IAEA symposia [1, 2]. Also, some examples are given in the Annexes to this report.

1.1. Objectives and scope of the guide

The principal objectives of this guide are (a) to identify the safety considerations in industrial radiotracing, (b) to provide guidance to competent authorities in converting ICRP recommendations into appropriate policies and regulations applicable to radiotracing, (c) to assist competent authorities in implementing an appropriate program for control of industrial radiotracing, and (d) to provide users with guidance in the planning and safe implementation of radiotracer investigations.

Industrial radiotracer investigations are usually conducted to obtain information about an industrial process and usually involve short-lived radioisotopes. The guide covers the safety aspects of the use of radiotracer investigations in industry, including that radioactivity which may be incorporated into consumers' products as a consequence of the investigation. The guide is not concerned with the application of radioactive tracers to geological, hydrological or agricultural investigations, although, inevitably, it will overlap with some of the earlier IAEA documents involving tracer applications conducted outside industrial process systems.

1.2. Basis of recommendations

The guide is based primarily upon the recommendations of the International Commission of Radiological Protection (ICRP) and published IAEA Safety Series. As such, revisions will be required as adjustments to international safety guidelines or advances in technology are made.   The provisions of the guide should be treated as recommendations which are generally applicable.

2.   BASIC RADIATION SAFETY CONSIDERATIONS

The objective of radiation protection is to limit radiation exposure so that the risk of harmful effects to the individual and to society is as small as possible compared with the benefits to be derived from the use of ionizing radiation.

2.1. Principles

Principles of protection are embodied in the recommendations of the ICRP [3] (Appendix I contains some recommendations of the ICRP concerning limitation of doses for exposures from controllable sources), upon which most international standards [4] and national codes and regulations are based.

The recommendations and guidance provided by international agencies embody some principles of particular importance to those planning to use radiotracers in industrial processes.   These are: (1) Unnecessary exposures should be avoided;   (2) Operational control should be provided so that the resulting doses are as low as readily achievable and are justified by the benefits;   (3) Compliance with the relevant dose limits (IAEA Safety Series No. 9) should be ensured; (4) The resulting dose to the whole population should be much smaller than the corresponding limit, which should be viewed as a sum of minimum necessary contributions and not as a permitted total apparently available for apportionment.

2.2. Consideration of radiation workers and members of the public

When considering how to achieve the lowest practicable radiation dose in radiotracer work, it is necessary to distinguish between exposures to radiation workers and to members of the public.   The radiation workers in a tracer investigation normally include those engaged in (a) the preparation of the tracer compound, (b) the introduction of the tracer into the process, (c) product sampling or process measurement, and (d) radiation protection duties.   In some instances, a few persons not directly involved in the investigation need to remain in the controlled area established at the tracer injection point to perform duties essential to plant operation.   Efforts should be made to minimize the number of such persons who should

10

be regarded as radiation workers. All other persons should be regarded as members of the public.

The radiation received by members of the public, if any, from radiotracer investigations will generally result from release of wastes in liquid or gaseous effluents, or by incorporation of the radioisotope in a product which subsequently becomes accessible to the public. Careful selection of the radioisotope and equipment with high detection efficiency play an important role in assuring that the dose to members of the public is maintained as low as practicable. The relative dose or dose commitment from a variety of radioisotopes feasible for any given application should serve as a final basis for radioisotope selection. Once a radioisotope has been selected, planning should include a specific determination that the smallest practicable amount of radioactivity will be used, released to the environment, or contained in products.

2.3. The critical pathway and critical groups

Reconcentration of the released radioactivity, along a pathway leading to man, should be taken into account in the dose assessment. Tracer investigations which could result in radioactive materials being incorporated in products that are intended for ingestion, inhalation or application to man[1], should not be conducted unless clearly justified by sufficient benefits to society. The decision as to whether such radiotracer investigations are permissible is one of national policy. Some countries have regulations which prohibit such use.

ICRP recommendations on dose limits for members of the public are intended to provide standards for design and operation of radiation sources so that it is unlikely that individuals in the public will receive more than a specified dose. The variability resulting from a range of personal and environmental parameters makes it impossible to determine with precision the maximum doses that might be received by the individual from a given operation. In practice it is feasible to take account of this variability by the selection of appropriate "critical groups" in the population. These small and relatively homogeneous groups should be representative of those individuals in the population expected to receive the highest doses, and the ICRP considers it appropriate to apply the relevant Dose

---

[1] Typical products that would be excluded by this requirement are foods, drugs, cosmetics, or their components. The term "products intended for application to man" usually refers to products which are applied to the skin (such as ointment) or are in close contact with the skin (such as jewellery).

Limit to the mean dose of these groups. As pointed out in ICRP Publication No.7, in some situations, for example in preliminary planning or when the dose to the critical group will clearly be very small, it may not be necessary to make the detailed studies required for the identification of the critical group and the assessment of its dose. It will then be convenient to postulate a hypothetical group of extreme characteristics. The estimated dose to this hypothetical group will thus provide an upper limit to the dose that any real critical group could possibly receive. In the case of several sources of exposures of the population, the ICRP does not recommend how fractions of the limits could be apportioned among the various contributions. However, ICRP does point out that no single type of population exposure should take up a disproportionate share of the total dose and any exposure must be justified by a resulting benefit.

3. DETERMINATION OF RADIATION DOSE

3.1. Dose to radiation workers

The dose received by radiation workers during implementation of a tracer investigation will normally result from external radiation. Tracer investigations should be planned so that radiation workers will not inhale or ingest radionuclides although the possibility for doing so exists under emergency or accident conditions. Estimates of the dose from external radiation should be made using conventional calculational methods [5] during the planning phase of the study, and workers should be individually monitored for exposure, in addition to measurement of radiation levels during the operational phase.

3.2. Dose to the public

The principal pathway by which the public can be exposed will normally be through inhalation or ingestion of radionuclides released during the course of the radiotracer investigation. Investigations should be designed so that the dose received by members of the public when exposed to external radiation is negligible. In relation to external radiation the Dose Limits recommended by ICRP correspond to Intake Limits by ingestion or inhalation for each individual radionuclide [4]. For radiation protection purposes it is convenient to define a Derived Working Limit for concentra-

tion, $(DWL_c)^2$, assuming an extreme hypothetical critical group
drinking or breathing undiluted effluent. Exposing this hypothetical
group to this Derived Working Limit (DWL) continuously for one
year would give a dose commitment equal to the relevant Dose Limit.
If the concentration of radionuclides in liquid and gaseous effluents
from a tracer investigation at the point of release from the controlled area is well within these values of uncontrolled areas, persons designing the tracer investigations would not ordinarily need to
calculate the expected radiation dose to members of the public. The
competent authority should assess the potential exposure to members
of the public within a selected population group expected to be most
highly exposed, taking into account the duration of the tracer investigation and other activities which might contribute to the total dose.
If it appears that such releases may cause exposure in excess of
a very small fraction of the total dose limit for members of the
public, the competent authority should impose special restrictions.

3.3. Incorporation of radionuclides in consumers' products and release of activity to environment

With respect to incorporation of radionuclides in products
which become available to members of the public, the concentrations
of the radionuclides expected in the products should be calculated
and where possible the calculations should be confirmed by measurement. If the concentrations of radionuclides in products do not
exceed the DWL_c values for air and water in uncontrolled areas[3]
and if reconcentration of the radionuclides is unlikely and the
products are not designed for inhalation, ingestion or application
to man, calculations of dose to members of the public would not
ordinarily be required. If these criteria are not met, the competent
authority should require identification of the group of individuals
expected to be most highly exposed and the average exposure of a
suitable sample of the most highly exposed group estimated.
The release of radioactivity to the environment or the incorporation of radionuclides in products as a result of radiotracer
investigation has been small, and is likely to be so in the foreseeable future, relative to other activities. The total genetic dose is,

---

[2] The DWL_c for a nuclide is given by $DWL_c = $ (Intake Limit)/V, where the Intake Limit
is the value given in Ref.[4] for inhalation or injection, as appropriate, and V is the annual intake
of air or water (7.3×10^9 ml and 8.03×10^5 ml respectively).

[3] Some national authorities have specific regulations pertaining to types and permissible
concentrations of radionuclides in consumer products as a result of this type of activity. For
example, USAEC regulations on the subject are contained in 10 CFR 30, para.30.70.

therefore, believed to be a small fraction of the limit recommended by ICRP [3]. The genetic dose need not ordinarily be taken into account during the planning or operation phase of a tracer investigation. However, the competent authority should require periodic reports of quantities of radionuclides distributed in products which, when taken together with reports of similar releases from other activities, serve as a basis for the national authority to maintain surveillance over both the genetic and somatic doses to a population[4].

4. ROLE OF COMPETENT AUTHORITY

The utilization of radioactive tracers in industry is usually subject to control by the competent authority [4]. In discharging its duties, the competent authority normally performs the following functions with regard to the utilization of radiotracers:

4.1. Develops policy and criteria

The competent authority develops a policy regarding the use or conditions of using radiotracers. The policy decision ordinarily includes benefit-cost judgements which will be a formal benefit-cost analysis in some instances or it will be implicit in the policy decision. The competent authority also provides criteria or issues regulations about the conditions which must be met by industry to comply with established policy in carrying out radiotracer investigations.

4.2. Grants authorization

In making a request to conduct a radiotracer investigation, the industry normally provides relevant information and possibly a safety analysis of the proposed investigation. The competent authority then determines whether the proposed investigation is likely to be in

---

[4] After a meeting in April 1971, the ICRP issued the following statement with regard to dose from consumer products: "The Commission noted the increasing use of a number of consumer products containing small amounts of radioactive material, and the contribution to the population dose that these taken together could make even though the dose from individual sources is at present extremely small. In considering the relevance of this to the dose limit for the population, the Commission emphasized the importance of national authorities assessing the contribution being made by these products, so that an effective means of control may be instituted. In this regard, the Commission wishes to draw attention to a publication of the European Nuclear Energy Agency (OECD/NEA) Ref.[6], as an example of a method by which the total individual and population doses from all consumer products may be subject to administrative control".

conformity with established policy and criteria on the basis of the information submitted before granting authorization. (A brief but not exhaustive listing of the information to be made available by industries in their request for authorization is given in Appendix II.)

4.3. Conducts inspection

The competent authority may conduct inspections to determine compliance with safety requirements. It also undertakes appropriate action if safety requirements are not being met. Records of the receipt, transfer, use and disposal of radioactive material and of exposure to personnel participating in the investigation should be available to the competent authority to conduct inspection.

4.4. Maintains surveillance over the population dose

5. PLANNING AND DESIGN

5.1. Planning

Once the objectives of a proposed investigation have been defined, it is necessary to select a suitable technique (radioactive or non-radioactive). If the radiotracer technique offers advantages, the particular method to be adopted has to be selected and each stage examined with regard to radiation safety. Careful technical planning at the outset will assist toward an effective investigation with the minimum radiation exposure. Discharge of radioactivity resulting from the investigation, or incorporation of radioactive materials into products must also receive careful consideration to ensure that it is kept to the lowest practicable level.

In the majority of investigations, the effect on the environment will be negligible owing to:

(a) the small quantities of activity involved;
(b) the short half-life of the radionuclides;
(c) the expected dilution coupled with limited period of discharge.

In these cases, the data required by the competent authority for review and approval will ordinarily be minimal, since it can readily be demonstrated that concentration of radioactivity at the point of release will be well within the DWL values. In others where this cannot be readily demonstrated, estimate of population dose

15

should be made (i. e. in relation to the "critical group" as defined in ICRP Publication No. 7, Ref. [7]).

Complete information on concentration of activity in liquid and gaseous effluents and in products, product dilution factors and details of the critical group may not be readily available during the planning stage. However, on the basis of known data or findings of previous radiotracer work of a similar character, laboratory tests or calculations based on conservative assumptions, it is possible to estimate the order of magnitude of concentration in effluent or material of interest.

5.2. Design

5.2.1. Introduction

In designing a radiotracer investigation, the following factors have to be considered:

(a) Assignment of responsibility
(b) Selection of method
(c) Experimental design
(d) Radiation safety and protection.

These factors are all interrelated and will thus have to be dealt with together.

5.2.2. Assignment of responsibility

Because of the nature of radiotracer work which is carried out in industrial process plant, or in the field, it is essential to clearly define responsibility. It is necessary to designate a person as project leader to be responsible for the investigation. There is also need to make a specific person responsible for radiation safety who may or may not be the project leader. The general training and experience requirements for personnel undertaking radiotracer work in industry are outlined in the Appendix III. In addition, background knowledge of industrial conditions and acquaintance with the specific plant operations are necessary.

5.2.3. Selection of method

The selection of method for a radiotracer investigation involves the choice of technique, radiotracer and instrumentation. Factors influencing selection include:

(a) Satisfactory behaviour of tracer: the basic requirement for a radiotracer is that it should accurately represent the component of interest in the system to be studied [1, 8].

(b) Nature and energy of the radiation emitted and efficiency with which it can be detected.

(c) Half-life (as short as possible) consistent with the duration of the investigation.

(d) Aspects relating to radiation safety and protection (see 4. 2. 5).

5.2.4. Experimental design

The experimental design which will yield the desired information in the most efficient way must be established before conducting the investigations. This involves:

(1) The prediction of tracer behaviour under process conditions and possible deviations from anticipated behaviour.

(2) "Dummy runs" to perfect operating procedures and to overcome potential trouble spots. An alternative is a preliminary experiment on the full scale system itself using a small fraction of the radiotracer activity to be employed in the actual application.

(3) Identifying critical steps in handling the radioactive material.

(4) Preparation of detailed plan for the implementation of the investigation including evaluation of the anticipated experimental data. For the evaluation of the total tracer requirement and for the estimation of radiation exposures, engineering information about the process is required. Such information includes approximate flow rates, vessel volumes, dimensions of plant units and the location of discharge points.

5.2.5. Radiation safety and protection

5.2.5.1. General principles

Radiation safety and protection aspects enter into all the steps of design dealt with above. The choice method determines the radiation dose inherent in the operation. The selection of tracer will be a compromise between various considerations as previously mentioned. The radiation dose will be determined by factors such as:

(1) Activity of tracer
(2) Nature and energy of emitted radiation
(3) Half-life
(4) Radiotoxicity
(5) Radiochemical purity.

The activity of the tracer should be kept as low as practicable by judicial selection of method, experimental design, and the use of detectors of high efficiency. In practice it is necessary to choose a radionuclide with a half-life consistent with the duration of the experiment. A short-lived radionuclide would increase the initial activity (with corresponding increase in dose during tracer preparation and inspection); the use of long-lived radionuclides, on the other hand, lengthens the period of time during which the potential for radiation exposure exists. A compromise in selection is therefore necessary. Consideration of factors as indicated above may establish that several radioisotopes will be satisfactory for the work. The one finally selected should be that giving rise to the least potential radiation dose while successfully achieving the purpose of the investigation.

5.2.5.2. Protective clothing and equipment

In the majority of industrial radiotracer investigations, protective clothing and equipment are required at the tracer preparation stage and during the course of tracer injection. This is usually of a simple nature consisting of personal protective clothing, such as gloves and coverall/overalls, which minimizes the possibility of bodily contamination; and handling equipment which minimizes exposure by distance. More detailed guidance on this subject is given in IAEA Safety Series No. 22 [9].

5.2.5.3. Survey and monitoring equipment

Survey and monitoring equipment for industrial radiotracer investigations are usually simple. It is necessary to provide appropriate instrumentation to (a) confirm the dose rates estimated for the controlled area, (b) determine where the boundaries of a controlled area should be established, and (c) search for residual contamination in the working area where the tracer was introduced into the system or elsewhere within the controlled area before the area is released for unrestricted access. Personnel monitoring devices for the radiation workers should be provided as may be appropriate for the particular radioisotope used.

5.2.5.4. Emergency procedure

An accident is an abnormal occurrence arising in an investigation and may include

18

(1) a spill of radiotracer
(2) contamination of personnel
(3) malfunction of process plant resulting in unanticipated tracer behaviour
(4) an unanticipated release of the radiotracer to the environment.

Possible accidents in the various steps of a particular tracer investigation should be identified beforehand, and the procedures to be followed in the event of such accidents should be defined. The approach in relation to industrial radiotracer investigations is similar to that defined in IAEA Safety Series Nos. 1, 6 and 32 [10-12]. Decontamination procedures are also dealt with in the same documents.

6. OPERATIONAL PROCEDURES

6.1. Preliminary manipulation of the tracer

The initial handling of the tracer will normally take place in a radiochemical laboratory which should already be equipped with appropriate tools, shielding, and storage [13]. Here, the radio-activity to be employed is measured and transferred to an appropriate unbreakable container or injection equipment. It is sometimes awkward to shield large injection equipment during transport, and, in such cases, the final transfer may be more conveniently accomplished in the field.

In-plant transport from the laboratory to the field should be made in a vehicle with sufficient shielding to keep the external dose rates suitably small and providing for accident protection. Transport outside the plant should conform to the transport regulations contained in IAEA Safety Series No. 6 or appropriate national regulations.

6.2. Introduction of the tracer into the system

In industrial investigations, the tracer is frequently introduced into pipes or other conduits through a valve and under pressure. The tracer is at its most concentrated form at the injection site. It is usually necessary to establish a controlled area from which all but participating radiation workers are excluded. The controlled area should be established in accordance with the provisions of the IAEA Safety Series No. 9 or in compliance with requirements specified by national authorities. The area may be marked off or limited with ropes and posted with warning signs. Provision should be made (as necessary) for radiation shielding. In the case of a tracer solution, provision should be made to contain any spillage e. g. by use of trays lined with absorbent material.

Prior to introducing the radiotracer into a stream or process, injection equipment should be leak-tested as appropriate or otherwise checked for proper functioning. In some cases, it may be desirable to carry out a trial injection using inactive material. It is good practice to monitor the injector afterwards to establish that the injection has been complete before dismantling it.

6.3. Measurement of the tracer within the system

After the tracer has been injected into the system, the dilution which usually follows is ordinarily sufficient to eliminate the need for all but the simplest radiation protection measures at a sampling or in situ detection point. Further dilutions typically occur as the stream mixes with other streams or enters a storage tank. In those cases where large dilutions do not automatically occur, consideration must be given to temporary restrictions on sampling or to precautions in sample handling. It is usually unnecessary to restrict the access of persons to any site other than the injection site.

6.4. Post-operational surveillance

Prior to disconnecting the injector from the process system, it should be confirmed that residual radiotracer levels in the injector itself or in the part of the process system to which it is connected (e. g. a valve) are negligibly small. If necessary, the system should be flushed through (as indicated in Section 6.2) to reduce contamination to acceptably small levels. Where this procedure is unsuccessful, the equipment should be isolated by barriers posted with appropriate warning notices, until special procedures have been developed to dismantle or decontaminate the equipment or until radioactive decay has reduced contamination sufficiently for free access to be permitted.

Process material (solid or liquid) which is being retained to permit reduction of activity by decay, shall be kept in isolation in controlled storage until such time as it is acceptable for disposal. In the event of an accident leading to the dispersal of radioactive material, the extent of the contaminated area must be determined and roped off, posted with warning signs and otherwise secured against unauthorized access until decontamination has been completed and the area declared safe by the person assigned responsibility for radiation safety.

REFERENCES

[1] INTERNATIONAL ATOMIC ENERGY AGENCY, Radioisotope Tracers in Industry and Geophysics (Proc. Symp. Prague, 1966), IAEA, Vienna (1967).
[2] INTERNATIONAL ATOMIC ENERGY AGENCY, Nuclear Techniques in the Basic Metal Industries (Proc. Symp. Helsinki, 1972), IAEA, Vienna (1973).
[3] INTERNATIONAL COMMISSION ON RADIOLOGICAL PROTECTION, Recommendations on Radiation Protection, ICRP Publication No. 9, Pergamon Press, Oxford (1966).
[4] INTERNATIONAL ATOMIC ENERGY AGENCY, Basic Safety Standards for Radiation Protection (1967 edition), Safety Series No. 9, IAEA, Vienna (1967).
[5] INTERNATIONAL ATOMIC ENERGY AGENCY, Safe Handling of Radioisotopes: Health Physics Addendum, Safety Series No. 2, IAEA, Vienna (1960).
[6] OECD NUCLEAR ENERGY AGENCY, Basic Approach for Safety Analysis and Control of Products Containing Radionuclides and Available to the General Public, June 1970.
[7] INTERNATIONAL COMMISSION ON RADIOLOGICAL PROTECTION, Principles of Environmental Monitoring Related to the Handling of Radioactive Materials, ICRP Publication No. 7, Pergamon Press, Oxford (1965).
[8] INTERNATIONAL ATOMIC ENERGY AGENCY, Industrial Radioisotope Economics, Technical Reports Series No. 40, IAEA, Vienna (1965).
[9] INTERNATIONAL ATOMIC ENERGY AGENCY, Respirators and Protective Clothing, Safety Series No. 22, IAEA, Vienna (1967).
[10] INTERNATIONAL ATOMIC ENERGY AGENCY, Safe Handling of Radioisotopes, Safety Series No. 1, IAEA, Vienna (1962).
[11] INTERNATIONAL ATOMIC ENERGY AGENCY, Regulations for the Safe Transport of Radioactive Materials − 1967 Edition, Safety Series No. 6, IAEA, Vienna (1967).
[12] INTERNATIONAL ATOMIC ENERGY AGENCY, Planning for the Handling of Radiation Accidents, Safety Series No. 32, IAEA, Vienna (1969).
[13] INTERNATIONAL ATOMIC ENERGY AGENCY, Manual on Safety Aspects of the Design and Equipment of Hot Laboratories, Safety Series No. 30, IAEA, Vienna (1969).

NOTE ON ICRP PUBLICATION No. 9

The following information is taken from ICRP Publication No. 9. Those paragraphs selected are for easy reference to recommendations of particular relevance to the use of radiotracers in industry. It is suggested, however, that for an adequate understanding of ICRP recommendations, the reader should refer directly to the reports.

GENERAL

As any exposure may involve some degree of risk, the Commission recommends that any unnecessary exposure be avoided, and that all doses be kept as low as is readily achievable, economic and social considerations being taken into account. It should be noted that the dose limits are intended for planning the design and operation of sources leading to foreseeable conditions of exposure; the setting of "action levels" for exposures from uncontrolled sources depends on other considerations.

EXPOSURE OF INDIVIDUALS

Occupational exposure

In any organ or tissue, the Dose Equivalent due to occupational exposure shall comprise that contributed by external and internal sources resulting from the circumstances imposed by the occupation. It shall not be held to include the dose from any medical exposure, from exposure to natural background radiation or from other exposures received by the individual as a member of the public. The Commission wishes to emphasize that "medical exposure" refers to the exposure of patients in the course of medical procedures and not to the exposure of the personnel conducting or incidentally associated with such procedures.

In practice, the problem of chief concern is chronic exposure either at low dose rates or by intermittent small doses at high dose rates. Under these conditions it is reasonable to assume that the dose accumulated over a period of years is the controlling factor in determining the risk, provided the intermittent doses are sufficiently small. The Commission believes that a period of one year is the

most reasonable length of time over which to assess accumulated
exposures, but that it is also necessary to limit the magnitude of
a single dose. The Commission therefore recommends that in
any one year the Maximum Permissible Doses should not be exceeded,
but that in a period of a quarter of a year up to one-half of the annual
Maximum Permissible Dose, or, for internal exposure, a dose
commitment resulting from an intake of a radionuclide equivalent
in amount to the intake for one half year at the Maximum Permissible
Concentration, may be accumulated in conformity with considerations
on additivity and multiple organ irradiation. The recommended
values for the quarterly quotas may be rounded upward to the next
whole number. If necessary, the quarterly quota may be received
as a single dose, but the Commission believes that it would be
undesirable for doses of this magnitude to be repeated at close
intervals.

Members of the public

The Maximum Permissible Doses that have been established
for occupational exposure are regarded as upper limits, and the
doses may have to be individually monitored and controlled to ensure
that the Maximum Permissible Doses are not exceeded. The dose
limitation for members of the public is a more theoretical concept,
intended to provide standards for the design and operation of radiation
sources so that it is unlikely that individuals in the public will receive
more than a specified dose. The effectiveness of this is checked
not by observing individuals but by assessments through sampling
procedures in the environment and statistical calculations, and by
a control of the sources from which the exposure is expected to
arise. For these reasons it is seldom meaningful to speak of
Maximum Permissible Doses for individual members of the public;
instead, the Commission recommends that the term Dose Limit
should be used in connection with limitation of the exposure of
members of the public.

In any organ or tissue the Dose Equivalent is the sum of the
Dose Equivalents contributed by both external and internal sources.
It shall not be held to include any exposure from natural background
radiation or medical procedures.

The annual Dose Limits for members of the public shall be one-
tenth of the corresponding annual occupational Maximum Permissible
Doses. The figures are given in Table I, which summarizes the dose
limits for individuals.

The basis for the limitation of exposures of members of the
public is the dose to the various body organs and not the derived

24

TABLE I. SUMMARY OF DOSE LIMITS FOR INDIVIDUALS

Organ or tissue	Maximum Permissible Doses for adults exposed in the course of their work (rems/yr)	Dose Limits for members of the public (rems/yr)
Gonads, red bone-marrow	5	0.5
Skin, bone, thyroid	30	3
Hands and forearms; feet and ankles	75	7.5
Other single organs	15	1.5

criteria by which the dose is controlled. The actual doses received by individuals will vary depending on factors such as differences in their age, size, metabolism, and customs, as well as variations in their environment. The variation resulting from these sources makes it impossible to determine the maximum doses that might be received individually. In practice, it is feasible to take account of these sources of variability by the selection of appropriate critical groups within the population, provided the critical group is small enough to be homogeneous with respect to age, diet and those aspects of behaviour that affect the doses received. Such a group should be representative of those individuals in the population expected to receive the highest dose, and the Commission believes that it will be reasonable to apply the appropriate Dose Limit for members of the public to the mean dose of this group. Because of the innate variability within an apparently homogeneous group, some members of the critical group will receive doses somewhat higher than the Dose Limit; however, at the very low levels of risk implied, it is likely to be of minor consequence to their health if the Dose Limit is marginally or even substantially exceeded.

In some situations, especially in the planning of proposed operations or installations, it may not be practicable to make the detailed studies necessary for the identification of the critical group. To allow for individual variability it will then be necessary to apply an operational "safety factor" to the derived concentration limits applicable to a member of the public. In previous publications the

Commission has suggested values for safety factors for environmental exposures to radionuclides. However, as the values to be recommended for such factors would vary over a wide range, depending on the particular circumstances, no generally applicable values are given in this report.

EXPOSURE OF POPULATIONS

General

The mean dose for whole populations is determined not only by the doses to individual members but also by the number of persons exposed. The main contributions, in addition to that from the natural background radiation, are at present made by the medical uses of radiation for diagnostic purposes, the increasing use of radioactive substances, and the release of radioactive material into the environment. Protection measures will include both reduction of individual doses and, wherever appropriate, limitation of the number of persons exposed.

GENETIC DOSE

Assessment of genetic dose

The genetic dose to a population is the dose which, if it were received by each person from conception to the mean age of child-bearing, would result in the same genetic burden to the whole population as do the actual doses received by the individuals.

The genetic dose to a population can be assessed as the annual genetically significant dose[5] multiplied by the mean age of child-bearing, which for the purpose of these recommendations is taken to be 30 years. The annual genetically significant dose to a population is the average of the individual gonad doses, weighted in each individual for the expected number of children conceived subsequent to the exposure.

---

[5] The reader is referred to the 1962 report of UNSCEAR (UN General Assembly Official Records: Seventeenth Session, Supplement No. 16 (A/5216)) for a detailed discussion of genetically significant dose.

26

Genetic dose limit

The Commission recommends that the genetic dose to the population should be kept to the minimum amount consistent with necessity and should certainly not exceed 5 rems from all sources additional to the dose from natural background radiation and from medical procedures. The contribution to genetic dose from medical procedures should be kept to the minimum value consistent with medical requirements.

The Commission wishes to point out that it is important to ensure that no single type of population exposure takes up a disproportionate share of the total. The way in which this is done will depend upon circumstances which may vary from country to country, and will be determined by national, economic and social considerations.

INFORMATION SUBMITTED TO COMPETENT AUTHORITY

The following points outline the type of information which is frequently required by the competent authority prior to granting authorization to conduct radiotracer investigations. This list is illustrative. It is not necessarily complete nor is all the information listed necessarily needed. This depends on matters such as previously established information on certain types of radiotracer investigations or previously established information about the competence of the investigators.

1. <u>Identification.</u> Name and address of investigator and location of investigations.

2. <u>Objectives.</u> A description of the radiotracer investigation. The reason for choosing a radiotracer technique and the rationale for the radionuclide and activity selected should be explained.

3. <u>Radioactive tracer specifications.</u> The radiotracer used, radionuclidic composition, total quantity used per investigation, and chemical and physical form of the radiotracer.

4. <u>Personnel.</u> The training and experience of the project leader and the persons responsible for radiation safety, and the instruction given to team members. (Name and address.)

5. <u>Facilities and equipment.</u> The facilities and equipment to be employed in the conduct of radiotracer investigations and equipment available in case of accidents.

6. <u>Operating and emergency procedures.</u> Procedures utilized by the investigation teams, including the following:

(a) Procedure for transporting the radiotracer
(b) Procedure for injection
(c) Procedure for sampling
(d) Procedure for flushing and decontamination of equipment
(e) Methods and occasions for conducting radiation surveys
(f) Utilization of personnel monitoring equipment and protective clothing
(g) Methods for restricting access to controlled areas

(h) Method for storage of radioactive material
(i) Method for disposal of wastes
(j) Method for inspection, maintenance and testing of equipment
(k) Measures to be taken in the event of an accident, including procedure for notification.

7. Hazards evaluation

(a) Maximum frequency for conducting repetitive investigations. (For inspection or surveillance purposes the competent authority may request notification of the dates of investigations.)

(b) An analysis of the design of the investigation and operating procedures which demonstrates that doses to workers will be within prescribed limits and as low as practicable; and that investigations are likely to be conducted in accordance with regulations and criteria prescribed by the competent authority.

(c) An estimate of the activity and concentration of radionuclides expected to be released and the conditions of release, in air and liquid effluent streams resulting from an investigation or series of investigations; as appropriate (see Section 3), identification of the critical group and an estimate of the average dose to individuals within the group.

(d) The chemical and physical form of the radionuclide as it is incorporated into products (if any), its concentration in the product, and the total activity to be incorporated into the product; the nature and use of the product and the total quantity of product involved; as appropriate (see Section 3), identification of the critical group and an estimate of the average dose to individuals within the group.

TRAINING AND EXPERIENCE REQUIREMENTS

Training and experience requirements for persons engaged in the use of radioisotopes for tracer investigations can be considered from two aspects: planning and implementation. Formal training equivalent to college level instruction in subjects related to radiation is ordinarily needed to plan a radiotracer investigation. This will involve selection of the isotope, writing of the procedures, and devising and evaluating controls adequate to protect workers and members of the public. Many organizations that wish to conduct radiotracer investigations may not have, nor need, an individual with this type of competence on their staff on a full-time basis. Consultants are often used.

Tracer investigations should be carried out by personnel who are trained in handling of radioactive materials. They should have a working knowledge of: (a) calculations basic to the use and measurement of radioactivity, (b) radioactivity measurements, standardization and monitoring techniques, (c) biological effects of ionizing radiation, (d) the principles and practices of radiation protection.

Of the order of 30 hours, or a one week's training course, can usually be considered sufficient for instruction in these radiation safety oriented subjects for the necessary level of competence. In addition, a period of training in practical radiotracer work by assignment to an experienced team is advisable before any personnel is given responsibility for tracer investigations. The supervisor of an experimental team should have previous experience in radioactive applications.

All personnel taking part in radiotracer work should be thoroughly familiar with the procedure they must follow in the event of accident. For technicians, the procedure will be straightforward, consisting primarily of action designed to contain the incident, making the affected area as safe as possible and promptly reporting the occurrence to the supervisor of the team. The supervisor should have sufficient knowledge and experience of radiation safety to take charge of the incident, defining actions to be taken to remove all hazards and, if necessary, taking specified action to summon such additional expert assistance as may be required.

It is desirable that training should include some practice in dealing with simulated accidents. All personnel should be familiar with the use of personal monitoring devices, radiation survey instruments and any protection clothing which may be required in carrying out radiotracer investigations.

EXAMPLES OF TRACER APPLICATIONS

Several specific examples of application of tracer techniques in various industrial branches are described in the following section. All of these investigations have been successfully carried out after being approved by the relevant competent authority. In all the quoted examples the information required could most efficiently be obtained by the use of radioactive tracers.

Most of the examples given refer to investigations of units or operations which were part of the process system in large-scale industrial plants. One example describes the study of the dispersion of industrial effluent, while another example demonstrates the use of a radioactive gas tracer for ventilation efficiency study in an industrial laboratory.

In all cases described the exposure of the public was very small, assessed in terms of the hypothetical critical group as discussed in Section 2.3 and Section 3.

The References cited in these Annexes are gathered together in one list at the end of Annex 9.

MEASUREMENT OF LIQUID FLOW RATE WITH A PULSED INJECTION OF GOLD-198 TRACER [1]

Procedure adopted

For this application, many radioisotopes are available in chemical forms readily soluble in aqueous systems and possessing convenient half-lives and gamma-ray energies for external detection in process systems. The system under study involved the flow of about 40 m^3/min of concentrated sulfuric acid in a large recirculating system.

^{198}Au was selected as the tracer. Its 2.7-day half-life was long enough to permit repetitive measurements over a period of several days, yet short enough to allow for reasonably rapid decay after addition to the process system. Because of the rapid recirculation of the acid in this closed system, other radioactive flow methods were not convenient; and the determination was based on timing the initial appearance of the tracer with radiation detectors at two points on the 50-cm diameter flow line following a rapid injection.

Initial handling in the laboratory

The tracer as received was counted for its activity and an appropriate quantity (10 mCi) added to a larger volume of water contained in a 500 ml unbreakable plastic bottle. These manipulations were carried out using typical radioactivity laboratory procedures employing simple, remote-handling tools, and appropriate shielding. The gamma radiation level is about 0.24 mR/h per mCi at 1 m, but the 0.41-MeV gamma rays are easily shielded with 2-3 cm of lead. Transfers of the solution were made with a glass pipette operated with a syringe to suck up and to discharge the solution.

Handling at the application site

The tracer solution was transported to the field within a lead shield. At the field site, the tracer solution was poured directly into the injector using long tongs to grasp the bottle. The injection system had been first completely assembled onto the plant line and pressure tested for leaks at its connection and valves, then partially disassembled for loading.

To carry out these operations, the tracer solution was kept in a lead shield until the final transfer. Once the injector was loaded, the injection was made as soon as possible. During this transfer and injection, the tracer was not shielded; however, the operations were completed rapidly (a few minutes at the most) and there was very little radiation exposure to the operator, as measured with dosimeters on the hands and the body. Ten mCi gave less than 2 mR to the hand and less than 1 mR to the body.

The completeness of the injection was verified by surveying the injector with a gamma-ray survey meter prior to dismantling the injector. Once in the plant system, the tracer was immediately diluted with a large volume of flow; there was no further handling hazard.

Radiological safety

The initial handling problems concerned the tracer in concentrated form where the principal requirement was to reduce gamma-ray exposure by shielding and distance. After the tracer was injected, there were no further external radiation exposure problems as the high initial concentration was rapidly reduced by dilution and dispersion.

The remaining radiological safety problems concerned internal contamination and discharge of the radioactivity to the environment. In the series of measurements described in [1], the highest concentration reached after the injection of 250 mCi was 0.003 μCi ^{198}Au per gram of sulfuric acid. This was substantially higher than the MPC, 5×10^{-5} μCi/g, for discharge of effluents to uncontrolled areas. However, none of this acid was released as an effluent. The spent acid was combined with other acid sources and was regenerated and re-used.

The hazardous nature of the acid itself had already established operating procedures in the plant which avoided direct contact with the acid by personnel. The relatively short half-life led to rapid decay.

MEASUREMENT OF STEAM FLOW RATE WITH A PULSED INJECTION OF HYDROGEN-3 TRACER [2]

Procedure adopted

For determining the flow of a condensable gas, such as steam which exists in two phases during the measurement, it is desirable to use an isotopic tracer; in this case, hydrogen, to avoid fractionation of the tracer between the two phases. The tracer in the form of liquid water is injected rapidly into the steam line; downstream a small side flow of steam is withdrawn and condensed to water. Samples are collected continuously so as to encompass the tracer wave. The determination is based on the total sample method.

^{3}H emits beta rays only. The very soft radiation (0.018 MeV) may be counted directly in the form of water with a liquid scintillation counter or in an internal gas counter after conversion to hydrogen or other suitable gases.

Initial handling in the laboratory

There is no external radiation exposure hazard from ^{3}H, but care must be exercised to avoid ingestion, inhalation, or absorption through the skin. The vapour pressure of water at ambient temperatures is high enough so that when concentrated solutions or tritiated water are handled, the skin of the hands and arms should be covered with gloves and operations should be conducted in a chemical fume hood.

Transfer of solutions was carried out with a pipette operated with a syringe; pouring was avoided. A radioactive gas monitor (tritium monitor) was used during the transfer. The inlet of the monitor hose was positioned at the level of the face. The quantity used for the measurement was loaded directly into a pressure injection vessel in the fume hood or diluted into a large volume of water (ca. several hundred ml) for later transfers in the field. These later transfers were done in a well ventilated area.

Handling at the application site

All fittings and connections were first leak tested under pressure, then the loaded injector was connected to the valve on the plant line and to a high pressure gas cylinder to inject the tracer.

Radiological safety

The initial handling problems concerned internal contamination by the tracer. In the event of an accident in which the tracer solution is spilled, dilution with large amounts of water should be carried out at once. If spilled on the skin or on gloves which have some permeability, the tracer solution should be washed off first, then the balance of the spill should be flushed away.

After the injection into the plant line, the high initial concentration was rapidly reduced to low concentrations. For example, using the data of Ref. [2], consider the measurement of a flow of 45 000 kg/h of superheated steam at 35 atm using 80 mCi of ^{3}H. At the downstream sampling point, nearly all the activity was found in the first 100-second sample. This gave an average concentration of $6 \times 10^{-2} \, \mu\mathrm{Ci/cm}^3$ over this short time interval. This concentration at the sampling point was already below the MPC for this isotope in controlled areas. Only a 20-fold greater dilution was necessary to reach the MPC for uncontrolled areas. This additional dilution was rapidly reached at greater distances downstream and by blending with other streams of steam or condensate.

A bioassay should be conducted on any individual suspected of receiving ^{3}H internally. This is most simply done by counting a urine specimen collected a few hours to a day after possible exposure using the same counter as for the samples. For example, a body burden as low as 0.1 μCi can be readily detected.

MEASUREMENT OF GAS FLOW RATE WITH A
PULSED INJECTION OF KRYPTON-85 TRACER [4]

Procedure adopted

The use of an inert gas tracer permits its application for gas
flows over a wide range of pressure, temperature, and chemical
conditions. ^{85}Kr (10 yr) and ^{133}Xe (5 d) are both readily available
and may be conveniently detected. In this example, ^{85}Kr was
selected because of the availability of high-efficiency, beta-ray
detection equipment convenient for continuous on-site counting.
Following injection into the plant line, a small side flow is withdrawn downstream, reduced to atmospheric pressure, and passed
through a counting chamber containing a beta-ray sensitive Geiger-
Müller tube. The determination is based on the total count method.

Initial handling in the laboratory

^{85}Kr decays about 99.5% by the emission of 0.7 MeV beta rays.
These are readily shielded, while the radiation level from the
gamma rays and bremsstrahlung is only about 1 mR/h per curie
at 1 metre. ^{85}Kr should be stored in a fume hood. Storage may
be in a pressurized metal container or at atmospheric pressure over
mercury in a glass burette. The beta radiation is stopped by the
metal or glass wall.

The presence of even a small amount of gamma rays makes it
convenient to determine the quantity to be employed by gamma-ray
counting. The quantity needed was transferred into an evacuated
glass tube and the tube then sealed. When immediate use was
planned, a rubber septum cap was satisfactory for sealing; and
the transfer was made through a hypodermic needle. The loading
of the sample was conducted without handling the tube directly.
When the tube was moved, it was handled with tongs or other long-
handled tools. All manipulations were conducted in a chemical
fume hood, and a radioactive gas monitor (tritium monitor) was in
operation at the face of the hood.

Handling at the application site

The glass tube containing the tracer was inserted into a high-
pressure injector just before use. The injector, which contained

a metal rod for breaking the glass, was closed; the tube was then
broken.   The injector was then connected to a valve on the plant
line and also to a cylinder of high pressure gas to flush the tracer
into the line.

The completeness of the injection was verified by surveying
the injector with a gamma-ray survey meter prior to dismantling
the injector.   Once in the system, the tracer was immediately
diluted with a large volume of plant gas; there was no further
handling hazard.

Radiological safety

The initial handling problems concerned handling the tracer
in concentrated form, avoiding inhalation, and reducing gamma-ray
exposure by distance and shielding.   The transfer was conducted
in a fume hood to reduce any possibility of inhalation of the tracer.
In the event of an accident, whereby the glass tube is broken in the
laboratory, the room should be evacuated at once and not re-entered
until the time calculated from the volume and ventilation rate of
the room shows that it is again permissible to enter.   A beta radiation survey meter or radioactive gas monitor should be used when
re-entering.

Downstream from the injection point, the high initial concentration was rapidly reduced to concentrations far below the MPC.
For example, using the data of Ref. [3], consider the measurement
of a flow of 3000 m^3/h with an injection of 70 mCi of ^{85}Kr.   The
tracer was injected slowly over several minutes to hold the maximum count rate to 100-200 counts/s on the Geiger-Müller tube.
Downstream it was observed that the tracer wave required 10 min
to pass.   The peak concentration was 4×10^{-4} μCi/cm^3 and the
average concentration over 10 min was 3×10^{-5} μCi/cm^3.   Hence,
within the flow line at the sampling point, the tracer concentration was already approximately equal to the MPC for this isotope
in controlled areas.   Additional dilution occurred at greater
distances downstream and by dilution with other gas streams.
This gas was then vented to the atmosphere.

MEASUREMENT OF LARGE GAS FLOW RATES USING XENON-133 [4]

Procedure adopted

This example concerns a requirement to measure large flow rates (approx. 100 000 m^3/h) of synthesis gas (CO, CO_2, H_2, Ar) feeding a complex of chemical manufacturing plants. An accuracy of ± 1% (or better) was requested in order to calibrate a specially designed orifice plate system installed in the main, data from which would be used subsequently for costing and plant efficiency evaluations. Absolute calibration at such rates was evidently impracticable; factors such as the complex gas composition coupled with the high accuracy requirement ruled out all possible methods with the exception of a radioisotope technique. Among the radioisotope methods, those depending on pulse injection of tracer were rejected as being insufficiently accurate for this example. Accordingly, a method based on the continuous injection of gaseous tracer (dilution principle) was developed, in which steady-state conditions could be maintained for periods up to 30 min for each individual test.

Of the three gaseous tracers readily available (^{41}Ar, ^{85}Kr, ^{133}Xe), ^{41}Ar was rejected as impractically short-lived. ^{85}Kr was considered less desirable than ^{133}Xe, because it is essentially a beta emitter and hence the assay would be subject to indeterminate errors due to moisture condensing from sample gas during counting. This error would not be significant if the assay were to be based on the electromagnetic radiation emitted from ^{133}Xe. Selection of ^{133}Xe was also supported by the recognition that the freezing point of xenon lent itself to a convenient method for extracting xenon from large gas samples by a simple refrigeration procedure based on liquid nitrogen. This procedure allowed xenon assay to be carried out under conditions of high efficiency in the well of a sodium iodide crystal, thereby assisting accuracy and reducing the amount of activity required in the flow measurements.

Initial handling in the laboratory

The principal laboratory procedure consisted of the preparation of the injection gas reservoir. For each on-site test, approximately 100 mCi of xenon-133 was required. This was obtained in a glass capsule, the seal of which could be broken with the capsule attached to a gas handling line, by means of an iron slug operated by a magnet external to the line. The complete handling apparatus was accommodated in a fume consistency "hood" or "cupboard". Prior to breakage, the capsule limb was immersed in liquid nitrogen to freeze the xenon. Inactive carrier xenon was mixed with the radioactive material and introduced into a pre-evacuated cylinder of 30 litre (s. t. p.) volume. The gas cylinder was subsequently pressured to 100 atm with nitrogen.

Handling at the application site

The tracer reservoir was coupled to the flow control and metering equipment at the injection site. A temporary controlled area was established around the injection site. Samples were drawn from the main some 250 m downstream from the injection point.

Radiological safety

During initial handling, the beta dose rate from the xenon was small due to attenuation by the glass walls of the handling line. The gamma dose received by the operator was also small ($\sim$ 1-2 mR) because of the speed with which the transfer operation was completed. Once in the gas cylinder, the external gamma radiation was less than 2 mR/h on the surface of the cylinder. Possible accidental release of the xenon during the initial handling operation (e. g. following leakage or breakage) was taken into account by mounting the apparatus in a fume hood.

On site, leakage was again the principal hazard, the specific activity of the injection gas being approx. 3×10^{-2} μCi/cm^3. This was guarded against by leak testing of the injection equipment prior to use.

Samples ($\sim$ 5 litre volume) from the main presented no hazard at the activity concentration of approximately 3×10^{-7} μCi/cm^3.

This low concentration presented no hazard in the chemical processing plants using the synthesis gas, which in any case introduced further dilution.

DISPERSION OF INDUSTRIAL EFFLUENT IN A RIVER USING A CONTINUOUS INJECTION OF BROMINE-82 TRACER [5]

Procedure adopted

The purpose of this study was to obtain information on the circulation and mixing processes governing the dispersion of effluent discharged from a large chemical manufacturing site to the estuary of a river where interfering substances present in the water precluded the use of dyes as tracers, and subsequently, to develop a mathematical model of the estuary.

In order to take sufficient account of tides, it was necessary to maintain the injection of tracer into the effluent for a minimum period of three days. Sampling at three depths was continued for ten days from the start of the injection, at eight stations along a 19-km length of the estuary. The effluent was essentially aqueous and discharged continuously through a single outlet channel.

Bromine-82 as KBr was selected as the most appropriate tracer: ^{82}Br has an excellent gamma yield (325%) of high mean energy enabling large samples (several litres) to be assayed with good efficiency with a sodium iodide scintillation counter; the half-life of 36 h ensured the rapid removal of residual radioactive material from the river; the chemical form, KBr, ensured that the tracer would remain in solution in salt water.

Handling at the application site

The scale and nature of this study called for special authorization by the relevant national authority as well as approval from local river and port authorities. The experimental procedure was designed to minimize handling operations at the injection site because of the very large amount of radioactivity involved (initially 300 Ci ^{82}Br). The injection reservoir comprised a stainless steel cylindrical vessel with a primary shield of 10 cm lead surrounded by sand bags, erected on a concrete apron adjacent to the effluent channel. The height of the vessel was chosen to be the same as that of the activity container, the latter being delivered from the isotope production unit as a special consignment. Several practice activity transfer operations were carried out using 3-m long tongs in a manner simulating the actual transfer operation. Provision was made to assist solution of the activity (consisting of several hundred pellets) by pneumatic agitation of the water in the injection reservoir. Injection was accomplished by means of a remotely controlled electric pump feeding the tracer solution through shielded metal capillary tubing to the surface of the effluent stream. Two reserve pumping systems were also installed. The injection vessel, lines and pumps were carefully checked for leakage before the activity was transferred.

Radiological safety

The chief potential hazard in this application concerned the possibility of an accident at the injection site. The injection site was isolated by barriers. Careful checking and rehearsal, particularly of the activity transfer operation, reduced the period in which the consignment was exposed (unshielded) to approximately 3 seconds and the maximum dose received by a member of the injection site team was 150 mrem. Hoses were available to deal with the remote possibility of leakage by flushing into the effluent stream. The injection site was manned continuously throughout the three-day injection period.

The initial concentration of activity in the injection reservoir was 8 mCi/cm^3. At the injection rate of 0.5 l/h, the maximum concentration of activity in the effluent stream prior to discharge from the manufacturing site was 8×10^{-4} μCi/cm^3. Additional dilution at the outfall reduced the maximum concentration to approximately 2×10^{-4} μCi/cm^3. (The maximum permissible concentration in drinking water for members of the public is 3×10^{-4} μCi/cm^3 — see ICRP Publication No. 2.) Because the maximum concentration quoted would persist for a short period only, and in view of the fact that the estuary is not a source of drinking water, the application was considered to present no hazard to members of the public.

MEASUREMENT OF MIXING IN A LARGE STIRRED TANK WITH GOLD-198 TRACER [6]

Procedure adopted

The time needed for complete mixing was to be determined in a very large tank holding approximately 11 000 m^3 of heavy fuel oil (Bunker C). Since this was a commercial product that would be sold after the study, it was necessary to use a tracer which had a short half-life, but which could be prepared in an oil-soluble form chemically stable in the oil at 86°C, the temperature in the tank.

The tracer choice was narrowed to several gamma-ray emitting isotopes with half-lives ranging up to 10 d. ^{198}Au was selected. Its 2.7-d half-life provided sufficient time for delivery from the source of supply, chemical conversion to an oil-soluble form, and use; while its relatively short half-life provided for rapid decay.

To obtain a high volume sensitivity for counting and therefore a lower tracer requirement, and to observe mixing at several points within the tank, the detectors would be suspended directly in the liquid in the tank. Because of the relatively high temperature, liquid-tight bundles, each containing four large Geiger-Müller tubes, were used as the detectors.

Initial handling in the laboratory

A large amount of radiotracer was required for this application. The initial amount handled was 1 Ci, and about 0.5 Ci was added to the tank. The manipulation of the tracer and its chemical conversion to the oil-soluble form were conducted in a lead-shielded fume hood. All transfers were made with remote-handling equipment. The chemical conversion was kept as simple as possible to minimize handling. The final product was diluted with oil in a 1-litre bottle for use.

Handling at the application site

The tracer solution was transported to the field site within a lead shield. The bottle was carried to the top of the tank at the end of a long pole, the cap was removed with a long-handled decapping tool, and the solution was rapidly poured into the tank. The bottle was then rinsed with a solvent and the rinses also added to the tank.

The radiation level from the unshielded bottle (0. 5 Ci) was about 100 mR/h at 1 m but only about 10 mR/h at the end of the handling pole. The period from removing the bottle from its shield until the contents were poured into the tank required only a few minutes, and the exposure to the person handling the tracer at the site was less than 5 mR. Once in the tank, there was no further handling hazard.

Radiological safety

The initial handling of the tracer required extreme care in avoiding radiation exposure and the possibility of an accidental spill because of the large amount of radioactivity involved. The use of long tools, adequate shielding, simple procedures, and well-rehearsed practice of all operations resulted in minimal exposure.

The final step involving the introduction of the tracer into the tank required handling without shielding, but the use of distance by means of long tools and of speed successfully reduced the exposure to low values.

The final safety problem concerned the ultimate disposal of the tracer. The concentration of tracer in the oil after complete mixing was 4.5×10^{-5} μCi/cm^3. This was just less than the MPC (5×10^{-5} μCi/cm^3) for discharge to effluents and well below the exempt concentration (5×10^{-4} μCi/cm^3) for consumer products permitted by the USAEC regulations (10 CFR 30, paragraph 30. 70). Hence, it was possible to release this material for sale without additional holding time for decay.

VENTILATION STUDY ON A PROCESS DEVELOPMENT LABORATORY USING KRYPTON-85 AND XENON-133 [7]

Procedure adopted

Ventilation studies are greatly assisted by the use of radio-tracers in that data are obtained immediately and continuously. ^{133}Xe and ^{85}Kr are usually the most suitable tracer gases for such measurements; both isotopes can be measured with high sensitivity and are of low radiotoxicity (see IAEA classification in Tech. Rep. Ser. No. 15, IAEA, Vienna, 1963).

In the present example, it was required to determine the ventilation rate and in particular, to search for "dead zones" in a process development laboratory in which toxic gases (e.g. carbon monoxide) were used in experimental work being conducted in high pressure reaction vessels. ^{85}Kr was employed to measure the rates of flow in the air intake and outlet ducts of the building by the pulse velocity method. ^{133}Xe was employed to simulate toxic gas release in selected parts of the building.

Initial handling in the laboratory

A small volume injector was charged with approximately 1 mCi ^{85}Kr.

A gas cylinder was charged with 25 mCi ^{133}Xe by the method described in Annex 4. All preparative work was conducted in a fume hood vented to the atmosphere through independent ducting.

Handling at the application site

The injection systems were isolated in temporary controlled areas established within the test laboratory, each system being checked beforehand for leakage.

Radiological safety

The external radiation prior to injection from both tracer gas containers was negligible, requiring only that non-radiation workers be kept away from their immediate vicinity. The dilution within the building and residence time of the ^{85}Kr pulses employed for duct flow determinations was such that no significant inhalation hazard was presented to workers within the building. Applications involving deliberate internal (inhalation) exposure to persons are very rare. In this example where ^{133}Xe was released at a low rate and its movement observed by installed detectors, the average concentration for several parts of the test schedule was $\sim 10^{-6}\,\mu Ci/cm^3$ and it was not considered necessary to take special precautions for tests at these concentrations. In other parts of the schedule, higher concentrations were employed and in these, especially where the exposure could not be predicted with confidence, tests were conducted with the building temporarily evacuated, until the continuously recorded detector responses indicated that the tracer had dispersed to concentrations of a radiologically insignificant level.

CONTINUOUS MEASUREMENT OF LINING WEAR IN A
STEEL FURNACE WITH IRIDIUM-192 TRACER [8, 9]

Procedure adopted

The rate of lining wear in an electric arc furnace was to be
measured in order to analyse the factors which fundamentally affect
it. The manufacturing process was not to be disturbed in any way
and there should be no restrictions imposed on the sale of the product
as a consequence of the measurement. It was therefore necessary
to use a tracer for the lining wear in such a form that it could be
easily assessed in steel samples taken for other purposes (routine
analyses). The concentration of tracer should not exceed
0.2 mCi/metric ton of steel, which is the maximum concentration
permitted in steel for β- and γ-emitters with a half-life shorter
than 1 year, according to the regulations laid down by the Swedish
radiation protection authorities.

The tracer should meet several physico-chemical requirements,
such as:

It should (a) be soluble in the molten steel
          (b) have a low affinity to oxygen
          (c) be inert with respect to all components of the
              refractory material.

These requirements would only be met by a noble metal or
alloy of high melting point. The final choice was a platinum-
iridium alloy, containing 10% iridium, in the form of a wire of
0.2 mm diameter. On irradiation it gave rise to a number of
short-lived isotopes of platinum, iridium and gold in addition to
iridium-192. After a sufficiently long ageing period, only
iridium-192 of 74-d half-life remained. Samples for ordinary
spectrography are taken routinely during processing. These
samples are made in the shape of cylinders of diameter 35 mm
and thickness 15 mm. The same samples were to be used for
activity measurements. The mass of a heat was about 50 metric tons
and the wear was estimated to be 10 mm/heat in the hot-spot area,
5 mm/heat in the vault area and 10 mm/heat in the bottom area.
The wear was to be measured separately in these three regions
during three campaigns in succession.

A scintillation detector equipped with a ϕ 75 mm $\times$ 75 mm NaI(Tl) crystal connected to a scaler was used for measuring the activity concentration in the samples.

Initial handling in the laboratory

Before irradiation 1.25 m of the wire was wound on an aluminium bobbin. One end of the wire was fixed in a notch on the bobbin. To the other end a 1-m long copper wire of the same diameter was soldered. The purpose of this arrangement was to minimize the risk for obtaining non-uniform specific activity along the wire after irradiation and to ensure simple and safe handling of the irradiated wire. Ten days after irradiation, which was the time necessary for the short-lived isotopes to decay, the bobbin was placed in a rig shielded by 30 mm of lead. The end of the copper wire was loosed from the bobbin by means of tweezers of 30 cm length and then attached to an inactive copper wire, which had before been pulled through a device designed for cutting the activated wire in segments of suitable length. The pulling device was driven by a synchronous motor and was effectively shielded by more than 20 mm of lead. By letting the motor run for a known time, a predetermined length of wire was pulled out. While it was being pulled out it was enclosed between two strips of adhesive tape. The wire was cut and transferred to a shielded transport container. Because of the tape enclosure the wire remained stretched, which was a requirement for further handling.

Handling at the application site

When the brick-laying had reached that part of the furnace which was to be investigated a tape-covered wire was placed in a groove on each brick to be tagged. The wire was lifted from the transport shield to the brick by means of a pair of tweezers. It was then immediately covered with mortar. The groove and hence the wire ran perpendicularly to the working surface of the brick. The wire ended a few millimetres below the surface. The brick was placed in its position in the lining, whereafter it was covered by a 6-mm lead sheet to protect the bricklayers against radiation during the rest of the laying. When the bricklaying was completed, the lead sheet was removed. The furnace was then ready for use.

Samples from the tagged furnace were measured for activity. A standard was prepared from samples taken from another furnace into which a known length of wire was thrown. By the use of known values for the mass of the heats in the furnace under study and in the furnace used for preparing the standard, respectively, accurate figures for the actual wear at all times during the campaign could be calculated.

Radiological safety

The activity of the wire was 0.04 mCi/mm and the wear in the hot-spot area, which was subjected to the largest erosion, was 20-50 mm/heat. Hence the highest specific activity in the steel produced was 0.04 mCi/metric ton, which corresponded to approximately 2000 counts/min in the measurement of a sample. The activity content was evidently well below the specified limit, while at the same time permitting the measurements to be made with sufficient precision.

During the initial handling of the active wire care must be taken to avoid exposure. The dose rate close to the unshielded wire was considerable but the use of shielding and long tools reduced the dose to the operator to very small values.

This was also the case for the handling at the application site.

The risk connected with active wire left in the furnace when the campaign is over also had to be taken into account. The best way of eliminating this risk was to use only so much wire that it would be consumed with certainty during the campaign.

INVESTIGATION OF WOOD-CHIP TRANSPORT IN A CONTINUOUS DIGESTER FOR PULP-MAKING [10]

Procedure adopted

The quality of the pulp produced in continuous digesting equipment depends on the wood-chip movement and cooking liquor flow through the equipment. For obtaining a high and uniform quality all chips should move along parallel vertical paths at uniform velocity, hence spending the same period of time in the digester.

In an experimental investigation Teflon chips containing a radioactive salt ($^{24}Na_2CO_3$) were used to study the chip transport through a continuous digester of 45 m height and 4.5 m diameter. ^{24}Na was selected as a tracer because of its penetrating gamma radiation and suitable half-life. The tracer chip should be measurable from outside the digester also if it moved along the axis of the digester, which corresponded to an absorber thickness of 2.25 m of chip/liquor and 43 mm of steel. To meet this condition, 50-100 mCi of ^{24}Na had to be used in each chip. The chip residence time in the digester was of the order of 3-6 hours depending on production rate. The only feasible way of introducing an activity of 100 mCi ^{24}Na in a volume of the size of a chip was to use hollow Teflon containers (dimensions 40 mm $\times$ 20 mm $\times$ 10 mm with a central cylindrical cavity) filled with the radioactive salt and subsequently sealed. Teflon was chosen due to its resistance to the process chemicals. The higher density of Teflon as compared to wood was not considered to lead to erroneous results as the chip bed was fairly compact.

One chip at a time was introduced with the chip feed and its passage past a number of preselected observation levels was registered by means of several scintillation detectors equally spaced around the digester.

By this method each chip could be accurately followed through the digester. At the observation levels the chips could be located with a radial accuracy of within $\pm$ 5 cm and an angular accuracy of within $\pm$ 5° in the digester cross-section with a corresponding timing precision of better than 30 seconds.

In all, six radioactive chips were used.

Initial handling in the laboratory

The Teflon chips to be used in the tracer investigation were
loaded with radioactive sodium carbonate at the isotope production
laboratory of the reactor centre where adequate manipulation equipment and facilities were provided for safe handling of the rather
large activities of ^{24}Na required for each chip (of the order of
200 mCi at the time of preparation).

It was not considered safe to irradiate the Teflon containers
after non-active sodium carbonate had been introduced because
of the risk of degradation of the Teflon which might impair its
mechanical and chemical resistance.

Handling at the application site

The Teflon chips were received in shielded transport containers.
Before a chip was introduced into the digester, a calibration was
made in a controlled area with the chip positioned 20 m from an
instrument which had previously been intercalibrated with all other
instruments used. Once this calibration had been made, the chip
position within the digester could be calculated from the instrument
readings.

All transfers during the calibration procedure were made with
long handling tools. After calibration, the chip was hoisted up in a
bucket to the digester feed level and quickly transferred to the chip
feed stream by means of long tongs. The dose to the person handling
the chip could be kept to less than 5 mrem.

Radiological safety

The initial handling of the tracer material was performed at the
reactor centre. Hence there was no exposure to the members of
the experimental team.

The handling at the industrial site involved only limited exposures
of the order of 5 mrem for each active chip run, since the manual
operations involved were simple and quick.

The minimum distance between any chip and the digester wall
was approximately 1 m in the actual investigation. If a chip had
been transported close to the digester wall the dose rate on the
outside would have been very high. This possibility was, however,
of no great concern from the point of view of radiation safety since
the digester was physically inaccessible except at the top and the
bottom. The detectors were in fact suspended from the structure
at the top of the digester.

When the Teflon chips passed out of the digester into the blow tank, they disintegrated due to the sudden pressure drop. The active material then dissolved rapidly in the blow tank liquor. The concentration of tracer in the liquor would be approximately 5×10^{-4} μCi/cm^3 which was somewhat higher than the MPC for discharge (2×10^{-4} μCi/cm^3).

The concentration was further reduced by additional dilution in the system, by decay and by blending with other effluent streams before final discharge, resulting in a concentration much below the MPC when the effluent reached the receiver.

The concentration of ^{24}Na in the pulp produced was calculated to be so low that there would be no restrictions on its further processing or release for sale.

REFERENCES TO ANNEXES

[1] HULL, D.E., FRIES, B.A., GILMORE, J.T., Acid circulation, volume, replacement, and entrainment measured in an alkylation plant with radiotracer, Int. J. Appl. Radiat. Isot. 16 19 (1965).

[2] FRIES, B.A., Steam-flow measurement by the total-sample method, Int. J. Appl. Radiat. Isot. 16 35 (1965).

[3] FRIES, B.A., Gas flow measurements by the total-count method, Int. J. Appl. Radiat. Isot. 13 277 (1962).

[4] JOHNSON, P., "Application of the dilution principle to the measurement of gas flow rates in large-scale chemical processes", Radioisotope Tracers in Industry and Geophysics (Proc. Symp. Prague, 1966), IAEA, Vienna (1967) 615.

[5] JOHNSON, P.E., private communication.

[6] FRIES, B.A., private communication.

[7] JOHNSON, P.E., private communication.

[8] MATTSSON, H., et al. (Isotopes Techniques, Inc., Stockholm, Sweden), private communication.

[9] BERGH, S., SANDBERG, H., STÅHL, N., Continuous measurement of lining wear in steel furnaces with radioisotopes, J. Met. 21 No.2 19 (1969).

[10] ERWALL, L.G., FORSBERG, H.G., LJUNGGREN, K., Industrial Isotope Techniques, Munksgaard, Copenhagen (1964).

LIST OF PANEL MEMBERS

CHAIRMAN

Cunningham, R. E.
Directorate of Licensing,
United States Atomic Energy
   Commission,
Washington, D. C. 20545,
United States of America

PANEL MEMBERS

Beninson, D.
National Atomic Energy Commission,
Manager for Health and Safety,
Avenida Libertador 8250,
Buenos Aires,
Argentina

Johnson, P.
Imperial Chemicals Industries Ltd.,
Petrochemicals Division,
Research and Development Department,
Billingham, Teeside,
United Kingdom

Ljunggren, K.
Isotope Techniques Laboratory,
Drottning Kristinas Väg 45-47,
S-11 4 28 Stockholm,
Sweden

Michon, G.
Département des radioéléments,
Centre de recherches nucléaires,
Saclay, B. P. No. 2,
91-Gif-sur-Yvette,
France

Nakazawa, K.
Director of Radiation Safety,
Atomic Energy Bureau,
Science and Technology Agency,
No. 1-2-2-chome, Kasumigaseki,
Chiyoda-Ku, Tokyo,
Japan

Subbaratnam, T.   Bhabha Atomic Research Centre,
                                   Health Physics Division, RHC Section,
                                   Bombay,
                                   India

Vagner, J.   Czechoslovak Atomic Energy
                                      Commission,
                                   Slezska 9, Praha 2,
                                   Czechoslovakia

WORLD HEALTH ORGANIZATION

Shalmon, E.   Environmental Pollution,
(Co-Scientific Secretary)   Division of Environmental Health,
                                   World Health Organization,
                                   1211 Geneva 27,
                                   Switzerland

OBSERVERS

Hamard, J.   Département de protection,
                                   Commissariat à l'énergie atomique,
                                   Paris 15e,
                                   France

Jahn, Dr.   Zentralabteilung Forschungsreaktoren
                                      der Kernforschungsanlage Jülich,
                                   517 Jülich,
                                   Federal Republic of Germany

Yuan, H.   International Atomic Energy Agency,
                                   Division of Research and Laboratories,
                                   Vienna

SCIENTIFIC SECRETARY

Bernardo, B.C.   International Atomic Energy Agency,
                                   Division of Nuclear Safety and
                                      Environmental Protection,
                                   Vienna

HOW TO ORDER IAEA PUBLICATIONS

Exclusive sales agents for IAEA publications, to whom all orders
and inquiries should be addressed, have been appointed
in the following countries:

UNITED KINGDOM	Her Majesty's Stationery Office, P.O. Box 569, London SE1 9NH
UNITED STATES OF AMERICA	UNIPUB, Inc., P.O. Box 433, New York, N.Y. 10016

In the following countries IAEA publications may be purchased from the
sales agents or booksellers listed or through your
major local booksellers. Payment can be made in local
currency or with UNESCO coupons.

ARGENTINA	Comisión Nacional de Energía Atómica, Avenida del Libertador 8250, Buenos Aires
AUSTRALIA	Hunter Publications, 58 A Gipps Street, Collingwood, Victoria 3066
BELGIUM	Office International de Librairie, 30, avenue Marnix, B-1050 Brussels
CANADA	Information Canada, 171 Slater Street, Ottawa, Ont. K 1 A OS 9
C.S.S.R.	S.N.T.L., Spálená 51, CS-11000 Prague
	Alfa, Publishers, Hurbanovo námestie 6, CS-80000 Bratislava
FRANCE	Office International de Documentation et Librairie, 48, rue Gay-Lussac, F-75005 Paris
HUNGARY	Kultura, Hungarian Trading Company for Books and Newspapers, P.O. Box 149, H-1011 Budapest 62
INDIA	Oxford Book and Stationery Comp., 17, Park Street, Calcutta 16
ISRAEL	Heiliger and Co., 3, Nathan Strauss Str., Jerusalem
ITALY	Libreria Scientifica, Dott. de Biasio Lucio "aeiou", Via Meravigli 16, I-20123 Milan
JAPAN	Maruzen Company, Ltd., P.O.Box 5050, 100-31 Tokyo International
NETHERLANDS	Marinus Nijhoff N.V., Lange Voorhout 9-11, P.O. Box 269, The Hague
PAKISTAN	Mirza Book Agency, 65, The Mall, P.O.Box 729, Lahore-3
POLAND	Ars Polona, Centrala Handlu Zagranicznego, Krakowskie Przedmiescie 7, Warsaw
ROMANIA	Cartimex, 3-5 13 Decembrie Street, P.O.Box 134-135, Bucarest
SOUTH AFRICA	Van Schaik's Bookstore, P.O.Box 724, Pretoria
	Universitas Books (Pty) Ltd., P.O.Box 1557, Pretoria
SPAIN	Nautrónica, S.A., Pérez Ayuso 16, Madrid-2
SWEDEN	C.E. Kritzes Kungl. Hovbokhandel, Fredsgatan 2, S-10307 Stockholm
U.S.S.R.	Mezhdunarodnaya Kniga, Smolenskaya-Sennaya 32-34, Moscow G-200
YUGOSLAVIA	Jugoslovenska Knjiga, Terazije 27, YU-11000 Belgrade

Orders from countries where sales agents have not yet been appointed and
requests for information should be addressed directly to:

Publishing Section,
International Atomic Energy Agency,
Kärntner Ring 11, P.O.Box 590, A-1011 Vienna, Austria